T0195636

Billionaire Success

JAVONTE' JENNINGS

authorHOUSE®

AuthorHouse™
1663 Liberty Drive
Bloomington, IN 47403
www.authorhouse.com
Phone: 1 (800) 839-8640

Published by AuthorHouse 01/19/2017

ISBN: 978-1-5246-5725-3 (sc)
ISBN: 978-1-5246-5724-6 (e)

Contents

THANKS TO

* GOD

* FRIENDS

* FAMILY

NAIL GUN

FRAMING NAILS

EYE PROTECTION

AIR COMPRESSOR

HEAD PROTECTION

2X4 STUD

$2.76 MAY INCLUDE TAXES

GARAGE

DOGHOUSE

SHED

"MENDING PLATES"

PETS WATER

CHICK STARTER

$10.00

CHICKEN FEED

$14.00

COMPUTER WITH INTERNET

GLOW LIGHTS

HAY BALES $7.00

MAY INCLUDE TAXES

REGULATOR

INVERTER

CAR BATTERY

RED!!!!! AWALYS GO ON THE (+) BLACK GOES ON THE (–)

CAR BATTERY 1 (KITCHEN)

RED!!!!! AWALYS GO ON THE (+) BLACK GOES ON THE (-)

CAR BATTERY 2 (DVD & TV)

RED!!!!! AWALYS GO ON THE (+) BLACK GOES ON THE (–)

CAR BATTERY 3
(COMPUTER & PRINTER)

RED!!!!! AWALYS GO ON THE (+) BLACK GOES ON THE (–)

CAR BATTERY 4 (BACK-UP)

RED!!!!! AWALYS GO ON THE (+) BLACK GOES ON THE (–)

CAR BATTERY 5 (BACK-UP)

RED!!!!! AWALYS GO ON THE (+) BLACK GOES ON THE (−)

RABBIT TV

SINK

TOILET

SHOWER HEAD

BLUE PVC PIPE

FOR COLD WATER

RED PVC PIPE

FOR HOT WATER

BACKHOE

VEGETABLE

SURVEY OF YOUR SELF

BREAKFAST

__ A.M – __ A.M

COMPUTER

__ A.M – __ P.M

PLAZA WORK

__ P.M – __ P.M

COOK DINNER

__ P.M – __ P.M

TV

__ P.M – __ P.M

☺END OF SURVEY☺

SOAPY WATER

CANDY RAPPERS

TIPS

BOXED MILK

RELISH PACKS

MUSTARD PACKS

MAYONNAISE PACKS

JELLY PACKS

SPRAY BUTTER

BUCKET PICKLES

STICK CHEESE

PACKAGED BACON

CANNED CHICKEN

CANNED SPAM

CANNED TUNA

CANNED BEEF

DRY FRUIT

POTATOES

YAMS

SOUPS & VEGETABLE CAN FOODS

PROPANE TANK

GRILL

WATER JUGS $6.00

WATER STORAGE

"THE HOME DEPOT" MEMEBERSHIP

STAPLES

WEED BURIRY

MULCH

PAPER PLATES

600 PLASTIC KNIVES

$10 MAY INCLUDE TAXES

600 PLASTIC FORKS

$10 MAY INCLUDE TAXES

600 PLASTIC SPOONS

$10 MAY INCLUDE TAXES

Printed in the United States
By Bookmasters